CORBY
IRON & STEEL
WORKS

To Dare

Best wishes

Steve

Purcell

Movements from the Blast Furnace area to the rest of The Works could be seen from the main road overbridge on Rockingham Road, near the West Gate. Here we can see D16 (English Electric D1049 of 1965) taking some ladles to the Steelmaking plant in April 1977.

CORBY
IRON & STEEL WORKS

Steve Purcell

TEMPUS

This scene on 1 January 1980 shows the original and famous Corby Candle. This flared off excess Coke Oven Gas (COG) with an orange flame, helping to keep the COG main at a constant pressure. The ignition system of this bleeder consisted of a coil of platinum wire that was connected to 110v and, when the power was switched on, it glowed like an electric fire element. It was situated close to the cone at the top and occasionally burned out, which meant a climb up the access ladder to the top platform. On a windy day, the top would sway – which was okay as long as eyes were kept on the job in hand. Looking at the surrounding buildings under these circumstances could induce feelings similar to seasickness!

First published 2002
Copyright © Steve Purcell, 2002

Tempus Publishing Limited
The Mill, Brimscombe Port,
Stroud, Gloucestershire, GL5 2QG
www.tempus-publishing.com

ISBN 0 7524 2769 5

TYPESETTING AND ORIGINATION BY
Tempus Publishing Limited
PRINTED IN GREAT BRITAIN BY
Midway Colour Print, Wiltshire

Contents

This statue was erected in front of the Civic Centre in August 1989 and is dedicated to all the ex-steelworkers from Corby's past.

Acknowledgements

This book would not have been possible had I not been given assistance from a number of sources, including my parents who nurtured in me the need to record history as it unfolds. In addition to my own photographs and diagrams, others were loaned by Mike Incles, Gordon White, Bob Freeman, Don Long, the late P.H. Wells, Ian Bateman and Reg Arnold, who provided the wonderful cover shots. Corus Steel (formerly British Steel Corporation) allowed me access to the roof of their administration block in order to take the 2001 views of the iron and steel works site and gave me permission to publish them, for which I must express my appreciation. Credit must also be given to my wife Sue, who not only made sense of the captions I had written, but also gave up the dining room table for six months so that I could store everything within easy reach. Encouragement was also forthcoming from a number of ex-Stewarts & Lloyds workers, to whom these photographs mean so much. Without all of the above, this labour of love to preserve The Works in this way would have taken even longer than twenty-two years to be published.

Steve Purcell, September 2002

Introduction

Much has been written about the rise and fall of the steel industry in the UK, but Corby holds a special place in my heart and, at the risk of repeating what has been said before, I would like to give a brief outline of its history.

Ironstone was first discovered and worked around Corby in Northamptonshire as long ago as Roman times, but not until the late nineteenth century did it become an important resource. The extent of the ore bed was unearthed by the excavations of the railway builders, especially between Leicester and Wellingborough. These discoveries led to small blast furnaces being built next to the ore beds in Kettering, Wellingborough, Cransley and Islip, and pig iron became an important source of revenue.

It was the construction of the Kettering to Manton railway in the mid-1870s that revealed huge deposits of Northamptonshire sand ironstone in the Corby area. In 1880, Mr Samuel Lloyd, of the Birmingham tube makers Lloyds & Lloyds, visited the area and his findings led to the formation, in that year, of the Cardigan Iron Ore Co. This venture leased the mineral rights from Lord Cardigan, the landowner, who lived at Deene Hall.

Lloyds Ironstone Co. was formed in 1885 to take over and expand the business and, by the turn of the century, three separate quarries were in operation. Each with its own railway system, the company owned six steam locomotives, also pioneering the use of a steam-powered digger to gain access to the iron ore.

The first two blast furnaces were built in 1910, on a site alongside the original quarries. The onset of the First World War saw a vast increase in production and a third furnace was built in 1917. The Lloyds Ironstone Co. was taken over by Alfred Hickman, iron and steel manufacturers of Bilston, in 1919 who were, in turn, taken over in 1920 by Stewarts & Lloyds.

Stewarts & Lloyds came into existence in 1903, upon the merger of the two largest tubemakers in the country, namely Stewarts & Menzies of Glasgow and Lloyds & Lloyds of Birmingham. In the years following their acquisition of Alfred Hickman, the Stewarts & Lloyds Board of Directors commissioned H.A. Brassert & Co. Ltd, consulting engineers of Chicago, Illinois, USA, to put forward a proposal to build an integrated iron, steel and tube works on a greenfield site at Corby. A scheme costing £3.3 million was accepted and construction of the new plant began in early 1933, with production commencing in 1934 when the new Bessemer plant came into operation.

Before 1933, Lloyds were using three blast furnaces and these were replaced by a plant of four larger furnaces that were commissioned at the rate of one per year from 1934. In 1938, No.1 Blast Furnace (BF) was taken out of use and rebuilt to the same dimensions as the new No.4, then brought back into operation in 1939. Other than No.3 BF, the furnace shells remained essentially the same until closure in 1980, with only the number of tuyeres increasing over the years. They all had a conventional two-skip, double-bell, McKee-type distributor charging system and the complete rebuild of No.3 BF, in 1974, into a three-bell, high top-pressure furnace, with its own Lurgi gas scrubber, was the only change in furnace outline during the life of the Stewarts & Lloyds plant. Each blast furnace had three stoves, with internal combustion chambers, that were 100ft high but the rebuild of No.3 BF entailed the top of Stoves 7,8 & 9 being raised another 30ft to accommodate the increase in internal volume of the new furnace.

The first Sinter Plant came into operation in 1934 and was the first in the country to use the Dwight-Lloyd process. Both Nos 1&2 Sinter Plants were removed from service in 1950 and demolished to make way for a new office/changehouse and canteen building, having been replaced by Nos 3&4. Eventually, two larger and more modern versions were built, the last being commissioned shortly before the rebuild of No.3 BF.

Coke was prepared in the Glebe plant that reached the maximum of 141 ovens in five batteries in 1953. The need for more coke led to the building of another battery of fifty-one ovens at the Deene site. Room for extension and other plant led to a large area being levelled and footings for a new blast furnace were put in before plans were shelved in 1962 because of a drop in steel demand worldwide. Nationalisation in July 1967 put a stop to these plans permanently and no further expansion ever took place.

Steelmaking began at Corby on 26 December 1934, in the newly erected basic Bessemer Plant. Initially, there were three 25-tonne capacity vessels installed and by 1941 there were two more in production. Trials using the Dickie process to control the nitrogen content of steel were conducted, followed by oxygen enrichment of the blast in 1947. The results of these trials were used to adapt the steelmaking process when BOC opened a tonnage oxygen plant in 1960, in redesigned 30-tonne capacity converters. Two 150-tonne Open Hearth Furnaces and their associated teeming bay were built in a new building to the north of the Hot Strip Mill and opened in 1949. Higher demand for steel brought about the decision to build a BOS plant, in an extension of the Open Hearth building, with three 100-tonne vessels capable of producing 1.5 million tonnes of steel per year and it was commissioned on 3 July 1965. In thirty-one years, the Bessemer produced 17,944,908 tonnes of steel, while the BOSC managed 11,149,239 tonnes in fifteen years. The Open Hearth Furnaces closed in 1971, having produced 4,394,625 tonnes of steel in twenty-two years.

During 1941, two electric arc furnaces and a teeming bay were added on to the end of the Bessemer building. These produced a different shape of ingot and were sent off to the Bilston works for further processing. They were also the last part of the plant to cease production in December 1980 when the last of the 2,324,652 tonnes of steel were cast.

The ingots produced in the steelmaking throughout the life span of the Corby Works were put into one of twenty soaking pits, where they were reheated until they were 1,320°C throughout. Once they had been sufficiently 'soaked', they were transported to the Heavy Rolling Mill and reduced into billets. From there, they were placed into one of the strip mill furnaces to be reheated for the last time in the Steelworks, then put through one of two strip mills and coiled for transportation to the Tubeworks for the final processing. Until 1950, they could only go into the Continuous Weld Plant, where four mills produced tubes ranging in size from 0.125in to 4in nominal bore. Over the years, various other tubemaking mills were built, including the Electric Resistance Weld (1in to 4.5in OD), Plug Mill (2in to 5.5in seamless) and the Extended Surface Tube plant, where helical fins were welded to the surface of the tubes, for heat exchangers.

My apprenticeship and subsequent qualification as an Instrument Mechanic gave me access to all parts of The Works and, over the next few years, I had the opportunity to take pictures inside various areas. Closure in 1980 was a massive blow to us all and many left the industry altogether, while others left the area completely. I was one of those who emigrated and did not witness the destruction that was wreaked on the site. Memories have kept The Works alive – such as one of the most memorable opening lines of any AUEW mass meeting. With all the tradesmen assembled in the old Lancs & Corby building, the union leader stood up and announced in broad Scots tone, 'Brothers. Certain allegations have been made … and we all know who the allegators are!'

With that speech in your head, I must now let you look at our pictures and see why iron and steelmaking at Corby meant so much to me and many others.

One
The Minerals Division

W1400 No.2 at Barns Close quarry in June 1975. This dragline was built in 1960 for the Cowthick quarry and was one of four Ransomes & Rapier machines used in the area around Corby.

Dragline No.70 (Bu 59529). This dragline was imported from the USA and erected during 1950-1951. It was a Bucyrus-Erie machine with a 217ft boom and a 19cu.yd bucket. It differed from the later Ransomes & Rapier machines in that the boom was constructed of angle iron rather than tube and it spent its entire working life in the Brookfield Cottage quarry, which almost surrounded Kirby Hall. Upon closure of the quarry system in 1980, it remained out of use until early 1983, when it was dismantled by J&F Construction and taken back to the USA. One of the machine's 19cu.yd buckets has been preserved at the East Carlton Heritage Centre.

Once the ironstone bed had been exposed by the dragline, it had to be broken up to enable the shovels to load it into railway wagons. This was done by drilling a grid of holes into the rock and, using explosive charges, the bed was shattered into more manageable pieces. Here, in Earlstrees quarry, exploding charges can be seen above the cables running across to the railway line, set off by the man in the foreground. The loading shovel can be seen working, while above it can be seen the factory of British Sealed Beams, now occupied by the Weetabix company. The quarry was closed in 1973.

This view is of Priors Hall quarry and wood, c.1970. The earlier parts of this quarry were restored and turned into a golf course in 1963 and one of the greens can be seen on the edge of the picture. The route of the railway, from the quarry through a cutting to the Steelworks, became the route of the A43 Weldon bypass in 1982.

Opposite: Dragline No.71 (R&R 2121). This was the first of the W1400 machines and was built in 1951 by Ransomes & Rapier, in partnership with Tubewrights, which was a subsidiary of Stewarts & Lloyds. Tubewrights fabricated the 282ft boom of tubular steel on site, which involved a considerable amount of on-site decision-making as nothing like it had ever been attempted before. No.71 was disposed in 1983 by R.M. Rees (Contractors) Ltd and moved to the USA to be used as spares for the W1800 that worked in the Oakley quarry.

Dragline W1400 No.2 (R&R 3072) was built in 1960 for use in the Cowthick quarry, near Stanion. At 303ft, the boom was the longest of all the walking draglines at Corby, while the 23cu.yd bucket was the same size as the one on *Sundew*, which was built four years earlier.

This is the view from the cab of W1400 No.2 as it slewed over to empty the 23cu.yd bucket. The bucket was pulled along the ground towards the cab to fill it and the tooth marks can be seen in front of the control panel below Frank Sprignall's right hand.

Tansley Huffer (left) and Frank Sprignall in front of W1400 No.2 prior to the 'walk' across the A43 in September 1971. The 1,700-tonne machine was carried on two 56-tonne feet, dragging a 1,000ft-long cable behind that provided the electrical supply to power the huge motors. Newspaper reports suggested at the time that there could have been a problem had Frank been the driver 'as he hasn't got a licence allowing him to drive on the road' – he got to work from Lowick each day on his bike!

W1400 No.2 began its working life in Cowthick quarry, but by 1971, the overburden there was so deep that only about 4,000 tonnes of iron ore could be produced per week. It was decided to open a new quarry called Barns Close on the other side of the A43 Kettering to Stamford road, where the overburden was much less and production of 10,000 tonnes per week was anticipated. Once safely across the A43, Tansley Huffer used the 23cu.yd bucket to remove the 6ft-thick layer of soil protecting the newly resurfaced road in minutes. The main road was only closed for about four hours and about 3,000 people watched the whole exercise.

Although not of the Stewarts & Lloyds era, this view of No.3 Conveyor in a quarry near Stanion Lane in 1919 has been included to show the long-standing use of steam machinery. It's hard to imagine that thirty years later this machine had evolved into the W1400.

Difficult to believe, but this 1980 view of Harringworth area quarries is now the site of the Rockingham Motor Speedway.

Dragline W1400 No.2 (R&R3072) in 1984. The last job this machine completed was to construct the final resting-place for the boom and it remained a visible reminder of the past, by the side of the re-routed A43, until H. Williams & Sons of Newport cut it up for scrap in 1987.

Dragline W1400 No.45 *Sundew* (R&R2825) in its final days at Shotley quarry, having built its own boom rest before being shut down for the last time. It was built in 1957 and named after the Grand National winner of that year and it lived up to its pedigree when it was walked from Exton Park quarry in the summer of 1974. The exercise of moving it to Shotley quarry was described as a stunt, but it was undertaken for sound reasons. The estimate to dismantle, move and re-erect it suggested a three-year period, whereas walking would only take three months. The compensation package for landowners, plus payments for services to the local authorities, would be a fraction of the dismantling costs alone. After meticulous planning of the route to give the minimum disturbance, *Sundew* set off at a speed of one mile every ten hours and had to cross ten roads, four rivers, a main railway line and seventy-four hedges on its twelve-mile trip. The journey lasted from 10 June until 5 August and required an extended stop halfway so that the six-mile cable could be transferred from Exton to Corby. The 282ft boom & 26cu.yd bucket remained thus until it was scrapped by Craylat Ltd in 1987, with the cab being preserved at the Rutland Railway Museum, not far from its birthplace.

Two face shovels at Shotley quarry in 1979. The smoking chimney in the background is that of No.6 Sinter Plant, which was built about the same time that No.3 BF was rebuilt.

The sheer size of Walking Dragline W1400 No.2 can be seen here as it dwarfs the Ford Transit van sitting nearby. The line of trees in the background shows the course of the main A43 road into Weldon, which is off to the left. At the time of this photograph, in 1975, the Minerals Division was responsible for producing around 3 million tonnes of ore per year, along with Corby & District Water Co., which supplied 7 million gallons of water to The Works every day. The Estates & Survey Department of the Minerals Division administered 26,000 acres of farm & woodland, as well as 70,000 acres of ironstone-controlled areas.

Opposite: Dragline W1800 No.81 (R&R3150) was the last of its kind to be built in 1963. This photograph was taken shortly after it began work in the new Oakley quarry in 1963 and shows the old A6003 that led from Oakley Hay into Corby. The end of the 282ft boom reached a speed of 23mph at full slew speed and the 36cu.yd bucket was the largest of the four R&R machines. The Works can be seen in the background, demonstrating the closeness of the quarry to both the town and the end user of the ore. The building under construction is the Strathclyde Hotel and the W1800 is sitting approximately where the Safeway supermarket has since been built. The fields across the road were turned into a housing estate in the 1970s. After the boom was lowered and rested in 1980, it was eventually dismantled by Craylat Ltd for R.M. Rees (Contractors) Ltd in 1983-1984 and shipped to the USA for use in the coalfields of Pennsylvania.

In 1959, an aerial ropeway was built from Rothwell West quarry to a terminal at the new quarry at Oakley from where the ore was taken by rail to the Corby Works. During its six years of operation, it carried over one million tonnes of ore, which would otherwise have been carried by road as the cost of extending the rail system far outweighed the £172,000 cost of the ropeway. This picture shows the Rothwell end as seen from the loading station.

Ironstone being loaded into wagons at Glendon quarry in 1976. The locomotive is No.20 (RR10273/68), one of three Rolls-Royce 6wDH diesels supplied new to Corby in 1968. They were sophisticated locomotives, with roller bearings, independently sprung wheels with flangeless centre drivers. Upon the closure of the system, it was sold to Ellis & Everard and put to work in their Bardon Hill quarry.

Two
The Blast Furnace
& Coke Oven Plants

This view was taken from Cottingham Road, and No.1 Quenching Tower (west) can be seen in full flow. The tall thin stack in the middle of the shot is the Blast Furnace Gas flare stack, showing that the flame could not be seen in daylight.

This 5-tonne steam crane, built by Thomas Smith of Rodley, was used for many years at the Glebe coke ovens to remove the damaged oven doors from the bench level and transport them to the repair area. This involved a trip through the West Quenching Tower, near where this picture was taken in October 1976.

Opposite: The main time office for the Blast Furnace and Sinter Plant area personnel was at the West Gate, with pedestrian access to Rockingham Road at the village end of the main railway overbridge. Once through the time office, there was a footbridge across the internal railway tracks, leading to pedestrian walkways to each area.

This stack was built shortly after the rebuild of No.3 BF to flare off the excess Blast Furnace Gas (BFG). It was the fourth & last bleeder on the BFG main, built especially to replace the other three that vented off raw BFG into the atmosphere. With a high carbon monoxide content, this gas was extremely dangerous in confined spaces and, in certain weather conditions, was blown into nearby Corby village, which in turn gave concerns for the well-being of the residents. The blue flame was visible only in poor light, but the roar could be heard at all times. The picture was taken in October 1976, just after it was lit for the first time.

This shot was taken of the Fuel Department personnel on the occasion of Arthur Skinner's retirement in 1978. The photograph was taken in the main control room and shows (from left to right) John Fuller, George Stoker, Bob Hayburn, John Loveday, Len Berry, Mel Dixon, Geoff Barratt, Geoff Kerfoot (partly hidden), Graham Wilson, Bob Binley (Fuel Dept engineer), Alan Tebbutt, Charlie Cowley, Arthur Skinner, Bob Freeman, Carl Beadsworth, Ray Keach, Harold York, Cyril Pascoe, Len Moss, John Kimmins and Les Rose.

This atmospheric shot, taken from the edge of the car park on Weldon Road, shows how different The Works looked at night. The West Quenching Tower was providing the steam when this picture was taken in December 1979.

Opposite: One of the many open-topped liquor collection pits in the Glebe Coke Ovens. Liquor was used to cool the hot gas in the collecting mains that ran along the top of the Battery to 90°C and any surplus to these requirements was pumped into these tanks for storage.

Another, almost aromatic, picture that shows the top of No.4 Battery at the Glebe Coke Ovens. The flame is coming from an open gooseneck on an oven that will shortly be emptied of coke.

The coke ovens were filled by hopper cars that ran along the Battery top and they received the prepared coal mixture from the massive concrete bunker that stood in the middle. It is from under the bunker that this May 1977 view was taken and it shows to good effect the reasons that respirators were required for those working on top for any length of time.

Each oven had gas valves on either side of the Battery which required greasing regularly and the operator is doing just that in the centre of this picture from May 1978. These valves were operated by the machine shown opposite. The tracks that can be seen on the right-hand side are those on which the Quench Car ran.

These are the gas flow and pressure instruments in the control room of No.5 Battery, Glebe Coke Ovens. As can be seen on the plaque above them, this section of the Glebe plant was built in 1953 and filled the remaining space in the line of ovens. The next Battery to be built was at the Deene site, which began production in 1961.

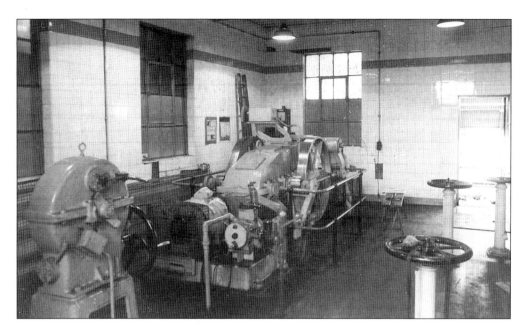

This is the control room of No.5 Battery at the Glebe Coke Ovens, with the reversing machine as the largest item in the room. By the time the plant closed in 1980, the total amount of coke produced at Corby was 26,234,000 tonnes in forty-seven years.

The reversing machine swung into action every half-hour or so, opening or closing a series of gas valves on either side of each oven. These valves provided fuel for the burning of the coal throughout the coking cycle, which was around twenty-two hours.

This is a view along the Coke Side of the Glebe Battery, with No.2 Quenching Tower (east) in the distance. Once the coke was quenched, it was tipped onto the wharf off to the left, then on to the blast furnaces by conveyor belt.

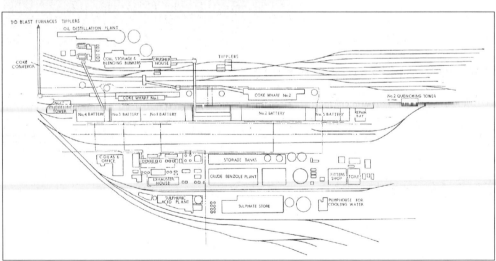

This plan of the Glebe Coke Ovens shows that, once No.5 Battery was built, there was no further room for expansion. Any future development would have to be elsewhere and, when it came, it was to be at the Deene site, which should have included a new blast furnace.

At the end of the coking cycle of each oven, doors were removed on both sides of the Battery. On one side, a large machine equipped with a ram, the end of which was of a smaller height and width than the oven itself, began pushing the red-hot coke into a large wagon called the Quench Car. Here we see the start of the push, with the coke oven shape being broken up by a chain across the guide. Once all was out of the oven, the car was pushed under the Quenching Tower and soaked for a certain amount of time, dependant on the requirements of the Blast Furnace. This produced clouds of steam that could be seen from miles away.

One of the first streets to be built near The Works was Stephenson Way. The fence at the back of the houses was at the edge of the railway line that runs from Glendon Junction to Manton, during the building of which iron ore was found in the area in the 1850s.

The iron came out of the tap-hole and ran into a trough, at the end of which was a dam and skimmer that produced a constant flow of iron down the runner and into the ladle. The arrangement can be seen here, along with the tap-hole drill on the right of the picture, and mud gun behind the furnaceman.

The iron ran from the dam into the 50-tonne ladles, passing along runners made of sand. This is No.3 Blast Furnace casthouse after the 1974 rebuild.

Opposite: At the end of the runner, a sand dam diverted the flow of iron over the edge of the casthouse floor and into the ladle. A platform was provided here for the furnaceman to stand on, in order to see when the ladle was full – when it was, the sand dam was knocked out and the iron then ran down to the next ladle point.

This view of the Ore Stockyard was taken during construction of No.2 BF in 1934. Once the Blast Furnace area had been developed, the area in the foreground became the Ore Preparation beds, where the freshly crushed ore was poured into a long pile to ensure an even consistency of material.

Opposite: Here we can see the complete run of iron from the furnace to the ladle. The blast furnaces produced a total of 30,879,525 tonnes of iron during the forty-seven-year life of the modern plant.

Once the iron had been tapped from the furnaces into the 50-tonne refractory-lined ladles, it was taken to the Steelmaking plant. There were four ladles available for three furnaces and five to the other when this photograph was taken in June 1964. Desulpherisation was carried out in the ladles with sodium carbonate (12lbs per tonne of hot metal) and lime (6lbs per tonne of hot metal) being added to them prior to casting, due to the acid burdening practice that was caused by using Northamptonshire ore. The locomotive is No 16 of the S&L fleet, built and supplied new by Hawthorn Leslie (3837) in 1934.

As well as iron being tapped from the furnaces, slag was also taken off at a separate tap-hole and run into these ladles. Here, S&L locomotive No.7 (Andrew Barclay 1268 of 1912) is removing the full ladles from the Blast Furnace area and will shortly be taking them down past No.2 Gasholder, under the Weldon Road down to the Tarmac plant.

The slag ladles were often assembled en masse in the sidings near the Glebe Coke Ovens and they made quite an artistic sight, as can be seen in this 1950s photograph.

Opposite: Blast furnace slag being tipped onto the cooling bank at the Tarmac plant. When cooled, the stone was broken up and made into hard-wearing roadstone, some of which was used to re-surface the roads inside the iron and steel works.

An outside view of No.2 Gasholder. Some of the graffiti can still be seen towards the top when this was taken in April 1980. As an indication of size, note the man walking at the base of the holder in the centre foreground.

Opposite: This view was taken looking straight up the inside of the collapsible piston access ladder, at the roof of No.2 Gasholder. The sides of the holder are covered in grease, which was used as a seal between the piston and the holder sides. The windows are in the roof of the holder.

Ken Croot (left) and Chris Bruce (right) of the instrument department, standing at the centre of the grounded piston of No.2 Gasholder, in May 1980. The extending access ladder platform is directly behind them as they lean on the ladder itself.

Opposite: This view is of the interior of No.2 Gasholder, showing plates and the piston-supporting frame. The collapsible ladder can be seen at the right of the picture, which was taken in May 1980.

This shot was taken after climbing the ladder – which took about ten minutes! An idea of the internal size can be obtained by finding Ken Croot and Chris Bruce, who are standing above the word 'NOT' at the centre of the picture.

Merv Collier (left) and Arthur Thomson (right) at the top of the access stairs for No.2 Gasholder, with the blast furnaces in the background. The chimney to the left of Merv Collier is the Blast Furnace flare stack that was not alight when this picture was taken in April 1980.

John McNeil (left) and Norman Evans (right), members of the gasholder maintenance crew, standing on the walkway across the top of No.2 Gasholder with The Works in the background. The picture was taken in April 1980.

This shot was taken looking straight down from the platform where Merv Collier and Arthur Thomson were standing in an earlier picture. The Blast Furnace Gas Main can be seen entering the gasholder towards the bottom left-hand corner and the lighter coloured pipe is a Natural Gas main. For size comparisons, note the car near the top of the picture, taken in April 1980.

This is a view of the Glebe Coke Ovens, taken from the platform at the top of No.2 Gasholder, taken in April 1980. The No.1 Quenching Tower is in the foreground and No. 3 Gasholder at the Deene is in the distance. To the right of the coke oven batteries can be seen the edge of the Glebe By-Products plant. The lighter coloured chimney to the left belongs to No.5 Sinter Plant, which was the last stack to be built in The Works in 1977. Only the water tower remains from this April 1980 scene.

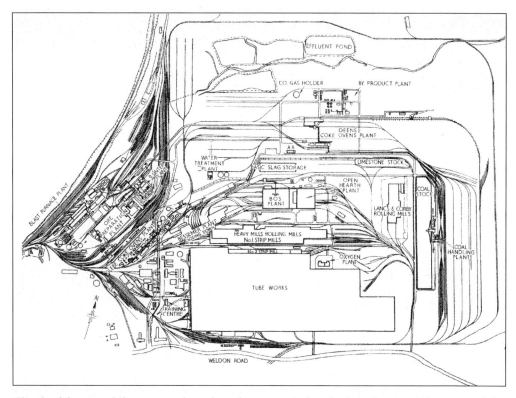

The final layout of the iron and steel works was completed upon the commissioning of the BOSC plant in 1965. This map was produced around that time and shows how lengthy movements of molten iron had to be in order to get from the blast furnaces to the west end of the Steelmaking building. The ingot output of the BOSC had to return towards the old Bessemer building and be delivered into the west end of the soaking pits, where the stripping bay was located. While steel was made in the Bessemer, movement of the ingots was relatively short as they came out of the south-east door and straight into the stripping bay.

Opposite: Sunlight shines about halfway down the collapsible ladder of No.2 Gasholder and shows how narrow the opening to climb down actually was – in fact, there were some people who could not fit inside and who had no choice but to treat it like a conventional ladder. Climbing down the outside when the piston was near the bottom of its travel was not for the faint-hearted.

The scene on New Year's Day, 1980, showing BR shunter 08 697 at the exchange sidings near the West Gate, with the Blast Furnace area in the background.

This shows how close the main road overbridge was to the exchange sidings throat. BR shunter 08 697 sits idle, as 1 January 1980 was the start of the industrial dispute that delayed The Works' closure.

Returning to the ore bed area, this shot shows the triangular cross-section of the crushed and mixed ore, with the Blast Furnace stockyard crane and Blast Furnace in the background.

After the rebuild of No.3 BF in 1974, the difference in capacity was quite marked, as can be seen here. The rebuild entailed making Stoves 7, 8 & 9 almost 35ft taller and erecting a small stack which was used to provide a back-draught for maintenance work on the tuyeres and their coolers.

One of the causes of air pollution from The Works was the emission of coal dust and gas from the Coke Ovens when they were pushed. Here, an oven on No.4 Battery at the Glebe site is showing signs of not being properly 'cooked' as the cloud of smoke envelops the area in 1977. The locomotive moving the Quench car is one of two Wellman Smith Owens used from 1934 until closure in 1980.

Opposite: The new Blast Furnace flare stack, as viewed from the No.1 Quenching Tower area in October 1976.

Outside view of No.1 Gasholder, which held 1 million cu. ft of Coke Oven Gas. The stairway to the top can be seen winding its way around the structure and, once inside the cupola on the roof, access to the piston was via the collapsible ladder that can be seen on the next page.

Inside view of No.1 Gasholder, with the piston access ladder fully extended down to the grounded piston. This view was taken in May 1977 when the holder was undergoing an overhaul. This meant that the entire 160 tonnes of concrete that was distributed around the piston had to be removed in order to make repairs. Great care and skill were required to replace the large slabs, to ensure that the piston moved up squarely when gas was introduced upon completion of the shutdown.

One of the last chimneys to be built in The Works was that of No.6 Sinter Plant, seen here from the Deene Coke Ovens office and amenity block in October 1977.

This is the Deene Coke Oven Instrument workshop in May 1977. The resident is Steve Purcell who was providing holiday cover for Alex McNeil, the area Instrument Mechanic. Note the 'pocket-sized' pipe wrenches hanging below the windows.

The By-Products plant at the Deene site had an acid recovery plant, which can be seen here in a view from the No.3 Gasholder.

Slag pots were moved from under the BOSC vessels and tipped over a bank from the tracks that can be seen in the foreground of this picture, taken in May 1977. Once the slag had cooled, it was broken up and taken to the Fison plant for conversion into fertilizer. The smoking stacks in the background show that two of the three vessels were making steel, as both precipitators are being used. These precipitators operated at 60kV DC with an efficiency of more than 99.6% being claimed.

Previous pages, left (page 64): This view was taken from the centre of the piston of No.3 Gasholder at the Deene site. The dark, grease-covered sides frame the roof-lights above the walkway from the outside freight elevator. This holder had a small, four-man elevator that provided access to the piston, rather than a collapsible ladder.

Previous pages, right (page 65): The size of No.3 Gasholder can be more readily appreciated in this shot, which was taken from the edge of the piston itself. As can be seen, the shape of this piston was domed rather than flat, as were the other two. John Sandy (left), Ken Croot (centre) and Chris Bruce (right) accentuate the 200ft height of the 2 million cu. ft vessel.

This view of the BOSC and Tubeworks was taken from the top of No.3 Gasholder, with the Deene Coke Ovens Battery and Quenching Tower in the foreground and the Coke Oven offices and amenity block to the right. The Ateelmaking building that was south of the Deene plant was originally built in the late 1940s for the Open Hearth plant and was later extended to accommodate the BOSC when it took over from the Bessemer Plant. When this photograph was taken in May 1980, the Dragline W1400 No.2 was in the Barns Close quarry at the top of Stanion Lane, next to the Weldon Stone quarry, and is visible beyond the BOSC.

Looking towards the Glebe Coke Ovens and Blast Furnace plant from the top of No.3 Gas-holder. The Battery top can be clearly seen, with No.1 Quenching Tower at the west end of the ovens. The sequence of Battery numbers could be a little confusing, being Nos 4, 3, 1, 2, 5 from west to east, and reflected the building sequence.

This is one of the last official aerial shots of the whole site, taken when No.3 BF was being rebuilt. Little else changed once the rebuild was finished and this shows the expanse of the iron and steel works, covering approximately two square miles. About all that remains of this scene now are the Tubeworks buildings that run horizontally in this photograph from 1974, along with the main Steelworks Administration Block.

A late afternoon shot on a sunny day highlights the Blast Furnace plant, with No.6 Sinter Plant chimney showing that production was underway. The steam rising above the water tower is coming from the valves on top of No.3 BF, as it is undergoing maintenance. This scene changed little from the 1930s until closure.

Brand new English Electric diesel (D913 of 1964), numbered D11 in the Stewarts & Lloyds fleet, is seen here moving some empty slag ladles past the high-pressure water tower of the Blast Furnace Gas cleaning plant, on its way back to the blast furnace.

Three
Steelmaking

One of the first buildings to be built for the new iron and steel works was the Main Stores, seen here in 1933, with the new No.1 BF being built in the background.

This shot of a Bessemer converter blowing shows the firework-type display that was caused by blowing oxygen-enriched air through 30 tonnes of molten iron and scrap. The man silhouetted is checking to see the stage of the blow, which he judged by the colour of the emissions. The Bessemer process was invented in 1866 and was the standard for 100 years, until the BOS process was perfected into the modern equivalent, the main difference being that pure oxygen is blown down on top of the molten material instead of through the bottom of the vessel.

This portrait of Bill Day was taken on the Bessemer platform, using only the light from the vessel under blowing conditions. The hats that the men used were made of felt and everyone always seemed to wear two of them. The glasses around his neck were for looking into the vessel to check on the liquid and the scarf prevented the heat and sparks from burning his neck and chest.

Overleaf: This picture of the Bessemer Plant is full of interesting snippets. The Mixer is pouring molten iron into a ladle while the overhead crane waits to pick up and discharge it into one of the vessels further to the right of the picture. There are a number of men in various places and poses, wearing bowler hats, safety hats or no hat at all. The size is difficult to gauge, but there is a 7-tonne ingot laying on its side below the Mixer near to the wagons full of bloom ends and other scrap, which will be picked up by the electromagnet nearby. This was taken in the late 1950s when the Bessemer Plant was in full production.

This shows how certain pieces of protective equipment have changed over the years. The safety helmet, seen here used in the Bessemer, was similar to those used in the oil industry, but the design changed so that there was only a peak at the front. The wheelbarrow was replaced by pallet-loads of additives that were in ingot form, small enough to throw into the vessel by hand. The Bessemer was also superseded by the BOS method, which was pioneered in the UK by Stewarts & Lloyds, Corby.

Another shot taken after closure – this shows the floor of the BOSC Plant, as viewed from the 110ft level. The ladles were those used for charging the vessels and, to assess the scale of the area, it is possible to see a man door in the centre of the picture.

There were two 35-tonne electric arc furnaces in an extension at the east end of the Bessemer building. The electric furnaces came under the Speciality Steel Division and continued in production until 21 December 1980. Here we can see one working on reducing the 100% scrap burden to a molten state.

Once the steel was made in the electric arc furnace, it was teemed in a similar fashion to that employed at the BOSC into ingots of a different shape. The round, stripped ingots were taken to other Works for rolling and further processing.

The Open Hearth Plant was built and commissioned in 1949. There were two Wellman Smith Owen furnaces, each with an annual capacity of 125,000 tonnes of high-grade steel. They were charged by a Wellman machine, commonly called 'The Tank' due to it having a long arm that resembled a gun. This arm picked up the bins full of scrap and additives and swung round through 180 degrees to put them into the furnace itself. The Open Hearth Furnaces were taken out of service in 1971, mainly due to the inefficiencies of the process when compared to the BOS method, which took less than an hour to make the same amount of steel.

At various times through the heat, temperatures were taken to ensure that the conditions were being maintained. Here we can see the long probe being manoeuvred into place in order to dip the thermocouple into the molten metal, with the protective shield keeping some of the heat generated by the process away from the operator. The slots on either side of the probe allowed the operator to see where the pyrometer was positioned.

The tapping hole was on the opposite side of the Open Hearth Furnace to the Tank machine. The process of producing 100 tonnes of steel took around eight hours, but did not require any liquid iron charge as the raw material was all scrap. Once the correct temperature of around 1,600°C had been achieved throughout the molten metal pool, the hole was opened and steel tapped into a large ladle, ready for teeming into ingots. The overflowing liquid is molten slag, which was taken out and poured onto the cooling bank at the north (or Deene Coke Oven) side of the building. Any steel that was taken off with the slag was reclaimed and put back into a vessel as scrap. This shot was taken in October 1961.

The Open Hearth building was enlarged in 1963-1965 to accommodate three 100-tonne BOS vessels, which were commissioned in July 1965. These replaced the Bessemer vessels as the primary steelmaking centre of Corby and remained until closure in 1980. This shot of No.2 Vessel was taken in 1966 as the molten iron from the Mixer was being charged. The control pulpit can be seen, from where the process was monitored on purpose-built instrumentation.

Another shot, this time of No.3 Vessel, being charged with molten iron in 1971. The charging crane was the same one used for the Mixer bay and was rated at 160-tonne capacity.

Temperatures of the molten metal in the BOS vessels were taken by a similar method as those employed at the Open Hearth. Here we can see the pyrometer being inserted in No.2 Vessel as the slag is being poured. The fumes are being sucked into the overhead duct by a 1,200hp electric fan unit, which took them to the Howden Lurgi fume extraction and cleaning plant before being released into the atmosphere.

To ensure the correct metallurgy of the blow (which is how each heat was known), a sample of the steel was taken, then sent off for analysis. The sample cooled quickly and was put into a small cylinder, which was then sent by Lamson tube to the laboratories nearby. This system involved putting the 'bullet' into a tube and propelling it along by compressed air, similar to the way that certain large department stores sent off payment for goods to a central cashier in their larger premises, before electronic cash tills were brought into general use.

When the blow was completed, the molten steel was poured into large ladles fitted with a plug. The Teemer operated this plug when the ladle was positioned above the funnel in the middle of four moulds, and it was opened by moving the lever on the ladle side. Here we can see the molten steel going into the funnel, which itself was linked to the bottom of each ingot mould.

Another shot of the teeming process, this time being supervised by two men. The ingots were cast from the bottom and the plug was replaced when all four moulds were full. The plug operating mechanism can be seen to good effect in this picture, along with the sparks coming from already poured moulds. They were then allowed to cool before the moulds were removed, which was sometimes done at the BOSC and at the Soaking Pit stripping bay at others.

From the BOSC, the ingots were taken to the Soaking Pits for re-heating. They were cast in fours on a railway carriage and we can see some being moved by D5 locomotive (North British 27409 of 1954) away from the Steelmaking plant. The ingots were, more often than not, taken to the stripping bay in their moulds, in order to preserve as much heat in them as possible. This meant less time was required to get them up to temperature, before they were sent to the Heavy Rolling Mill.

Opposite: This is a scene at the Soaking Pit stripping bay, with Bessemer ingots being prepared for their trip into a soaking pit. These ingots had rounded tops whilst the BOSC ones were square.

This is how the Soaking Pits, Electric Arc Furnace and the Glebe plant looked from the top of the BOSC building in 1980.

This view of the Lime Kilns was taken in 1980 after closure. Lime was brought from Derbyshire, as well as from the Weldon quarry situated at the edge of the area worked by W1400 No.2, and processed here for use in the steelmaking process.

Above: Once the ingot had been re-heated to a temperature of 1,320°C throughout, it was removed by the charging crane and put onto a chariot for transportation to the 48in Heavy Rolling Mill. Here we can see one of the 7-tonne ingots of metal being taken out, with the others still visible in the bottom of No.3 Soaking Pit.

Right: The charging crane is about to put the ingot from No.4 Soaking Pit into the chariot that can be seen at the top of this photograph of April 1980. There were twenty pits, each of which held ten ingots of 7 tonnes apiece and were fired by Blast Furnace Gas, Natural Gas or oil, and soaked at 1,320°C so that the heat was uniform throughout. The ingots were then withdrawn and sent to the 48in Heavy Rolling Mill.

From the soaking pit, the ingot was taken by chariot to the 48in Heavy Rolling Mill. This shot was taken from the roller's cabin and the ingot can be seen being removed from the pit by the crane and will be placed in the chariot, on the left of centre. The ingot was rolled back and forth until it was reduced into a bloom of the right size to go either to the 36in or 24in Mill.

Opposite: There were two shapes of soaking pit at Corby – round and square. The round ones were generally the newest and here we can see an ingot being removed from No.7 Soaking Pit by the large overhead crane.

This is the view as the ingot was reduced in shape at the 48in Heavy Rolling Mill, as seen from the roller's cabin. This mill was driven by a 7,500hp electric motor and the ingot turned after every few passes, to control the width and ensure uniform metallurgical quality.

The 24in Tandem Mill finished slabs from the 34in Mill to the size required by No.1 Hot Strip Mill (HSM), by reducing thickness in two stands and width in one edger. The finished sizes ranged from 6.5in x 4in to 21in x 5in.

The Fire and Ambulance Department of Stewarts & Lloyds were often entered into competitions against other forces across the region. They were quite successful, with the line-up of trophies here showing how successful they were in 1953 alone.

This was how the old No.1 HSM looked before it was rebuilt in the 1950s.

There were three slab re-heating furnaces of 110-tonne capacity, each being fired by Coke Oven Gas, Natural Gas or oil. This view of No.3 Furnace, No.1 Hot Strip Mill was taken from the Instrument Department workshop and is framed by the overhead crane and silhouetted by the sun shining through the roof.

The Hot Strip Mill had a waste heat boiler and its control panel is shown here after closure in June 1980.

Sunlight shining through the roof of No.1 Hot Strip Mill shows to good effect the steam being produced by the de-scaling water spray from No.1 Stand, while the outline of No.3 Re-Heat Furnace can be seen at left centre. This was the scene on Monday 19 May 1980, the last day of operation of the integrated iron and steel works at Corby.

In No.1 Hot Strip Mill, the bloom was converted into strip by passing it through a twelve-stand mill and, after leaving the last stand, the strip was conveyed along a roller path to one of two down-coilers in tandem. Here we can see Bob Kyle watching the progress of the strip along the length of the mill.

Once the strip had been coiled, it was placed on a cooling rack after being turned through 90°C. Here we see the coil as it is about to be lowered onto the metallic conveyor, which moved slowly enough so that the steel was not red hot when it reached the other end. It was then put onto a road train and delivered straight into the Tubeworks for further processing.

This view, taken from the coiler, shows the full length of No.1 Hot Strip Mill. A total of 32,535,736 tonnes of steel were rolled between 1933 and 1980.

Taken at 0830 on 1 January 1980, this shot from the Rockingham Road railway bridge reflects the mood of the day. It was the first day of the national strike against BSC by the ISTC union, who eventually settled for a 15.5% payrise. The strike gave a temporary reprieve to the iron and steel works, because the stocks for the Tubeworks had to be replenished, so the life of the plant was extended by a month.

Taken in May 1966, the view from the top of No.2 Gasholder shows the main Administration Block to the left and BOC to the right. Between them lies the main Steelworks car park with a large number of cars present, reason enough for a security cabin to be manned around-the-clock. It is possible to see the pedestrian bridge across the railway tracks in the left centre of the picture and the Tarmac plant is just visible in the distance.

There were railway sidings between the Glebe Coke Ovens and the Ore Crusher, where the coal wagons were marshalled. The chimney stacks are those of Nos 3, 4 & 5 Sinter Plants, with the crusher building dominating the left of this photograph, taken in May 1966. The shot was taken from the top of No.5 Battery, with the edge of No.2 Quenching Tower just visible on the right-hand side. The coal wagons, brought from the coalfields of South Wales, Yorkshire and the East Midlands, were emptied in 35-tonne side tipplers that turned them almost upside down.

Taken from the top of the Lime Kilns, the complete Glebe plant is shown here in May 1966. With the ore crusher buildings on the right of frame, the wagon sidings are full and the cooling towers are in full operation. The water tower remains in 2002, but nothing else in this picture survived demolition after closure.

Looking towards the Deene Coke Oven site from the top of the Lime Kilns, the piles are of slag brought from the BOSC and they form the raw materials for the Fison basic slag fertiliser plant. The final product was packaged into sacks and despatched countrywide by road and rail transport. The square brick-built building was a water treatment plant and the gasholder in the distance is to the left of the Deene By-Products plant. Another shot taken in May 1966.

This portrait of Jimmy Lindsay was taken in the Instrument Department mess room in May 1977, with the tea lockers behind him.

Opposite: The Coke Oven Gas flare stack at the Deene site had a pilot light to bring in the main flame, unlike the original Corby Candle. Here, Carl Beadsworth of the Fuel Department can be seen draining the liquor from the pilot light gas supply line. Most of the sludge that came out was made up of tar, naphthalene and other by-products that condensed and fell to the bottom of any section of pipe, preventing any gas from passing through.

After the official closure announcement was made, various parts of the plant were daubed with graffiti. This message was written on the cycle shed in the main steelwork car park and was visible from the Weldon Road. The steps to the right of the shed are those up to the pedestrian bridge that crossed the railway tracks down to the Tarmac plant.

An aerial view of the Eye Brook reservoir, which supplied most of the seven million gallons of water required each day by The Works. An Act of Parliament was required to build it and construction of the 1,600ft earthen dam was started in 1937, took three years to complete and the resulting reservoir covers just over 400 acres. Today, it is a centre for trout fishing.

This watchtower was where Security personnel could sit and guard the main Steelworks car park. The old brick-built cabin can be seen in this shot, taken from the cycle shed in May 1980. The gasholder in the background was also the scene of a large graffiti sign that could be seen from the town centre in late 1979.

Taken from the top of No.2 Gasholder in May 1966, both quenching towers are in operation and steam from No.1 obscures the batteries. The buildings to the left of the five Coke Oven chimneys are the ore crusher and Sinter Plants, with the ore beds in front of them.

Minerals locomotive D62 (ex-BR Class 14 D9515) is seen here heading towards the tarmac plant in May 1980. After being put into store near the old stripping bay in the Steelworks in December 1980, it was sent to Hunslett's for conversion to 5ft 6in gauge in November 1981.

Viewed from the Steelworks car park security tower, D23 of the Steelworks locomotive fleet is seen going down to the tarmac plant. This English Electric-built diesel (5354) was the first of five locomotives bought in 1971 and numbered in sequence with their maker's works numbers. Stewarts & Lloyds bought and numbered twenty diesel locomotives before being nationalised in 1967 – the first one being built by Crompton Parkinson in 1945. The British Steel Corporation bought a further sixteen, the last in 1975 and, without exception, all the Steelworks locomotives were put up for sale, then scrapped when none were sold, after the Tubeworks had picked the best for their stock.

Also viewed from the security tower, here we can see D64 (ex-BR D9549/65) returning from the Tarmac plant. It too was sent to Hunslett's in late 1981 for conversion to 5ft 6in gauge and exported to Spain. Of the twenty-four ex-BR locomotives bought for the Minerals Division in 1968, three were disposed of in the same manner, five went to be preserved at Heritage Railways due to their BR history and the rest were scrapped at one time or another before the end of 1982.

Here we see some of the instrument department crew on their last day at work, 15 May 1980. From left to right: Alex McNeil, Chris Bruce, Pete McKay, Ray Ward and Ken Croot.

Stewarts & Lloyds bought a LEO computer in April 1958 and housed it in a building that became known as the pay office, near to where the buses picked up the workers at the end of their shift. To quote from an official publication: 'It has approximately 4,000 valves and 10,000 resistors. The operational time is taken up as follows: pay roll for approximately 8,000 employees – 4 to 15 hours a week; warehousing invoicing (calculation and extension of approximately 600 invoices a day) – 45 minutes a day; maintenance and repairs, stores control and re-ordering – 40 minutes a day.' This is the keyboard of LEO – what a difference to modern computers!

Previous page: This view, of the entire iron and steel works site and the old village of Corby, was taken on 24 May 1980, when The Works was finally closed and before any major demolition of the site. The sheer size of the plant can be seen here, with the iron and steel works covering an area of approximately two miles by one mile, including almost 100 miles of railway lines within its boundaries. Demolition began in August 1980 when No.1 Coke Oven Gasholder – the middle of the three large holders in this shot – was felled by explosive charges. Within three years the site was cleared altogether, with very few buildings selected to survive the purging of Corby's skyline. Twenty-two years after this picture was taken, the scene has changed dramatically, with retail and commercial buildings occupying the Glebe Blast Furnace and Coke Oven sites, while the Rockingham Motor Speedway takes up most of the fields beyond the No.3 Coke Oven Gasholder at the Deene plant.

Four

Then & Now

The Works displayed two name boards in prominent locations, one on the side of the Blast Furnace Gas cleaning plant and this one at the main Steelworks entrance on Weldon Road. This sign remained until the company was nationalised in 1967 and the concrete footings can still be found in the undergrowth today, thirty-five years later.

These are the tracks that led down to Tarmac in May 1980. The footbridge led from the main Steelworks car park to the main entrance road. Molten slag from the blast furnaces was taken down to Tarmac to be cooled and finally made into roadstone.

This overgrown gully bears little or no resemblance to the previous view. The pedestrian bridge from the car park was removed and the cutting filled in to give a walkway across towards the administration block once the railway tracks had been removed. Picture taken in March 2001.

This shot, taken in 1958, shows how the approach to the iron and steel works main entrance looked before the administration block was built in the 1960s. The framework for the Stewarts & Lloyds sign can be seen in centre frame.

Since the previous picture was taken, the seven-storey administration block has been built, along with the training centre that can be seen to the left of centre. Picture taken in May 1980.

All the following pictures were taken from the top of the main Steelworks Administration Block in October 1977 and March 2001. This view is of the ironmaking site, with No.2 Blast Furnace Gasholder on the right-hand side. The main locomotive shed and wagon shops are in centre shot, with No.5 Sinter Plant chimney being formed towards the right of shot and it is one of fourteen stacks that can be seen in this picture of October 1977.

The same view, taken in March 2001. The only remaining building from 1977 is the old locomotive shed, which was the only way to landmark the original picture. Looking closely at this picture, it is possible to spot the water tower, which was almost hidden from view when the plant was working!

A similar view of the ironworks, but showing the path to the Steelworks car park. The blast furnaces are outlined against the skyline and the forming frame can be seen better on the top of No.5 Sinter Plant chimney. Picture taken in October 1977.

The paths at the bottom of the picture and the old locomotive shed were the landmarks. The trees have been removed and new ones planted and, when looking around the undergrowth to the right of centre, it is possible to find the concrete footings of the Stewarts & Lloyds sign. Picture taken in March 2001.

This shows the steelmaking part of the site, with both Coke Oven Gasholders visible. The single-storey building at the bottom of the picture was part of the DR&TD complex. The building to the right of centre is the Training Centre, which was opened by Sir Christopher Cockerell, inventor of the hovercraft. Picture taken in October 1977.

Twenty-four years later, the Training Centre is the only remaining building from the 1977 shot, other than the Tubeworks mills. The water tower is quite obvious in this view, but was almost hidden in the earlier picture.

The DR&TD complex is in the bottom of this October 1977 view, with W1400 No.2 on the horizon. The large electrical sub-station on the opposite side of the dual carriageway supplied The Works.

Quite a change can be seen in the intervening twenty-four years, with most of the DR&TD building making way for the inevitable car park, while the electrical sub-station has been replaced by part of the new Corby waste water treatment plant. The large area of railway sidings has also disappeared and the site of the former Barn's Close quarry is now a distribution depot for Peugeot vehicles.

The extent of the electrical sub-station can be seen here, along with a United Counties double-decker bus going through the only set of traffic lights in Corby. Picture taken in October 1977.

The Peugeot site can be seen here, along with the waste ground that once supplied electricity to The Works. The traffic lights remain the only ones in Corby. Picture taken in March 2001.

This shot, taken in 1968 from the DR&TD building, shows the locomotive shed to the left, with almost a dozen steam locomotives outside. The long building behind the locomotives is the main engineering shop and behind that is the main Glebe Coke Oven coal bunker and three of the plant's brick chimneys. In the distance are the chimneys for Nos 3,4 & 5 Sinter Plants and, to the right of the water tower and Corby Candle, is the old Bessemer building. The line-up of buses, on the right edge of the picture, shows that it is almost shift-change time. The odd-shaped brick-built building in the foreground controlled the railway crossing into the Tubeworks, with the Ford Anglia turning into the Tubeworks car park. On the other side of the road leading up to the Steelworks main gate is the main electrical sub-station and it is interesting to note that only the locomotive shed, water tower and roads remain of this scene and even the photograph vantage point has been demolished.